# 19th Annual Freight and Logistics Symposium

## Supply-Chain and Freight Challenges and Opportunities in a Changing Public and Private Environment

Summary Report & Presentations



Minneapolis, Minnesota, USA
2 December 2016

**Editor:**

**Michael McCarthy**

ISBN: 978-1-5108-4633-3

**Printed from e-media with permission by:**

Curran Associates, Inc.
57 Morehouse Lane
Red Hook, NY 12571

**Some format issues inherent in the e-media version may also appear in this print version.**



Printed by Curran Associates, Inc. (2017)

For permission requests, please contact University of Minnesota Center for Transportation Studies at the address below.

University of Minnesota Center for Transportation Studies
200 Transportation & Safety Building
511 Washington Ave SE
Minneapolis, MN 55455
USA

Phone: 612-626-1077
Fax: 612-625-6381

cts@umn.edu

**Additional copies of this publication are available from:**

Curran Associates, Inc.
57 Morehouse Lane
Red Hook, NY 12571 USA
Phone: 845-758-0400
Fax: 845-758-2633
Email: curran@proceedings.com
Web: www.proceedings.com

# TABLE OF CONTENTS

## 19th Annual Freight and Logistics Symposium

# Supply-Chain and Freight Challenges and Opportunities in a Changing Public and Private Environment

A Summary Report | December 2, 2016 | Minneapolis, Minnesota

**Sponsored by:**
Center for Transportation Studies

**Facilitated by:**
Continuing Professional Education,
University of Minnesota

CENTER FOR
TRANSPORTATION STUDIES
UNIVERSITY OF MINNESOTA

**In cooperation with:**
Minnesota Department of Transportation

Minnesota Freight Advisory Committee

Council of Supply Chain Management Professionals–Twin Cities Roundtable

Metropolitan Council

Transportation Club of Minneapolis and St. Paul

**19th Annual Freight and Logistics Symposium**

# Supply-Chain and Freight Challenges and Opportunities in a Changing Public and Private Environment

### *The Value of Reliable Data in Effective Supply-Chain Decision Making*

- Chuck Clowdis, Managing Director–Transportation, IHS Economics and Country Risk, IHS Markit

### *Panel: Minnesota's Freight Competitiveness in Our Region*

Moderator: Justin Johnson, Vice President of Sales, Bay & Bay Transportation

Participants:

- Bruce Abbe, Executive Director, Midwest Shippers Association
- Jim Carver, Director of Transportation, Land O'Lakes, Inc.
- Patrick Murray, Director of Supply Chain and Procurement, Cambria Company
- George Schember, Vice President of Transportation and Logistics, Cargill

### *What Can We Expect in 2017 in Light of the Election Results?*

- Jason Craig, Director of Government Affairs, C.H. Robinson
- Margaret Donahoe, Executive Director, Minnesota Transportation Alliance

**Center for Transportation Studies**
200 Transportation and Safety Building
511 Washington Avenue S.E., Minneapolis, MN 55455-0375
612-626-1077 | cts@umn.edu | cts.umn.edu

**Editor:** Michael McCarthy **Writer:** Darlene Gorrill **Designer:** Angela Kronebusch
**Photos:** Shutterstock

## Freight on the Move

How does the ability to move freight affect the economic health of a state, region, and even a city? How are the supply chains of businesses impacted by freight flow? And what challenges and opportunities does Minnesota face when it comes to leveraging and strengthening its freight modes?

The 2016 symposium offered a thoughtful examination and discussion of those questions and explored other topics that relate to improved mobility in the state, including congestion, regulation, and labor shortages, and the value of all freight modes to the state's economy. The symposium brought together representatives from the public and private sectors to gain a better understanding of freight flow issues.

The first presentation focused on the power of freight flow data in attracting industry to a location and ways to use data in making a compelling case for businesses to invest. Four industry experts from diverse organizations that depend on reliable freight movement highlighted key concerns in the second session, a panel Q&A. The symposium ended with a closer look at recent election results and potential changes that may result. The pages that follow offer a summary of those sessions.

### More information

Download this document and related materials from the 19th Annual Freight and Logistics Symposium: **cts.umn.edu/events/freight/2016**

## Decision Driver

### *The Value of Reliable Data in Effective Supply-Chain Decision Making*

With its roots in agriculture, Alabama may seem like an unlikely home for auto manufacturers. Yet in reality, manufacturing accounts for a significant share of the state's economy, with auto plants as healthy contributors.

The decision by auto giants such as Toyota, Honda, Mercedes-Benz, and Hyundai to locate their plants in Alabama was no accident, but instead it is a tale about the power of data in making the case for economic development.

Symposium speaker Chuck Clowdis pointed to Alabama as a state that actively pursued auto industry facilities by skillfully using data to show its strengths. As managing director for transportation with IHS Markit's economics and country risk sector, Clowdis helps connect public and private organizations to data that maximizes their opportunities for success—whether that involves finding the optimal location for facilities, better leveraging supply chains, identifying future prospects, or attracting businesses, such as those car manufacturers in Alabama.

"If you are wanting to attract investors, if you are wanting to attract that person who wants to put a plant or needs to put a plant in the Upper Midwest to your county or city in Minnesota, then you need reliable data," Clowdis says.

Presenting the data in a clear, compelling way helps all supporters to share the story behind the numbers. One state uses a baseball-card-like summary.

Source: IHS Markit

"[It] says, 'Here are attributes of my county,'" he says. "'Here's what we produce; here's what we import; here is the proximity to an intermodal ramp, for example; here are the number of truck lines, the number of railroads, the number of draymen, the closest international airport.' ... It is empowering, because at the end of the day what we want to do is attract jobs to counties."

Different types of data are available from diverse sources, Clowdis says, such as the federal government, state agencies, planning organizations, and commercial sources.

One key resource, the Freight Analysis Framework, integrates data from a variety of sources to create a comprehensive picture of freight movement among states and major metropolitan areas by all modes of transportation.

IHS offers Transearch, which compiles 1.365 billion bits of data from public and private sources that can show multimodal goods movement; U.S. freight flow information; commodity by volume, mode, and lane; truck flow sample data; and supply-chain details, among other information, at national, state, county, city, and even zip-code levels.

Chattanooga, Tennessee, Clowdis' hometown, turned to IHS for data to help convince Volkswagen to locate a plant in the city. The site selection firm wanted to know the proximity of other automotive suppliers and the state of the local labor force. The data helped show that the location worked well for Volkswagen's needs.

Minnesota—with companies that are global in scope and have strengths in many modes—is poised to adapt to changes that are coming, and, in the case of e-commerce, changes that are already here.

"There are compelling stories to be told about Minnesota. When you tell them with the support of good data, good visuals, and sustainable data that is reliable, then you can tell the story a lot better," Clowdis says, "and you can convince that company to put that plant here."

"If you are wanting to attract investors, if you are wanting to attract that person who wants to put a plant or needs to put a plant in the Upper Midwest to your county or city in Minnesota, then you need reliable data."
— Chuck Clowdis

## Challenges, Issues, and Strengths

***Minnesota's Freight Competitiveness in Our Region***

A featured panel discussion provided a closer look at the advantages, disadvantages, and opportunities of moving freight as Minnesota-based organizations.

**Panelists**

**Justin Johnson** (panel moderator), vice president of sales at Bay and Bay Transportation, posed a series of questions to the participants. With 15 years of experience in sales and operations, Johnson currently manages all sales for each of three business units of Bay and Bay.

**Bruce Abbe**, executive director of the Midwest Shippers Association since 2008, previously worked as a consultant and provided professional management services in agricultural policy, health care, and communications.

**Jim Carver**, director of transportation for Land O'Lakes, Inc., has worked in both the carrier and shipper segments in a variety of roles. He currently provides leadership to transportation team members, private fleet management, supplier and technology selection and implementation, and capital budgeting.

**Patrick Murray**, director of supply chain and procurement with Cambria Company, has worked as a professional in operations and supply-chain management for more than 20 years. He currently manages Cambria's global raw materials planning, strategic sourcing, procurement, logistics, warehousing, and production-fulfillment activities.

**George Schember**, vice president of transportation and logistics at Cargill, is responsible for transportation and logistics activities across Cargill's four global regions and multiple modal groups. He also is the process lead for the managed-goods movement and storage area of Cargill's process organization.

## Q: What does your organization do, and what is its impact on the Minnesota market?

**Bruce Abbe:** The Midwest Shippers Association, established by the Minnesota Legislature in the early 2000s to help spur growth in the specialty grains industry, works to keep its sector globally competitive and capture opportunities. Gradually the organization transitioned into a trade association that covers a five-state region and includes 100 members. A number of its members produce identity-preserved grains with specific traits that require careful handling by plants and transportation by container to global markets.

"This whole area with new food safety regulations coming on across the world, this need for identity preservation, for traceability, is flowing into agriculture on a continual basis," Abbe says. "So containerized shipping is really, really critical."

**Jim Carver:** Land O'Lakes began as the Minnesota Cooperative Creameries in 1920, and has evolved into a true agricultural co-op with more than 10,000 members. About 90 percent of its product is shipped domestically, with private fleets that haul milk and carrier partners. The company does business in Canada, Mexico, and Europe, and is expanding into Eastern Africa and South Africa.

Land O'Lakes also works with other companies that are also sometimes competitors. "The agricultural world really is very integrated," Carver says, "and it's very much based on partnerships and relationships. Those are very important to all of us at Land O'Lakes."

**George Schember:** Cargill operates in many arenas, including food, agriculture, trade, and industrial and financial products. "Generally, I would say we move a lot of stuff," Schember says. "We joke about ourselves as 'Tons-Are-Us.'... We ship 45 million metric tons of product to ourselves, which is a different way of thinking about it."

Some main food segments for Cargill include cocoa and chocolate, sweeteners, and oils. On the industrial side, the company supplies bio-based transformer oils and ethanol in places such as India. It also is a large red meat producer in the United States and a very large poultry producer outside the country. In addition, Cargill provides feed for about one-third of farm-raised salmon globally.

**Patrick Murray:** Cambria started 70 years ago in the dairy business but moved into the natural quartz stone countertop business in 1999. The company has a one-million-square-foot facility in Le Sueur, Minnesota.

"We don't manufacture anywhere else in the world," Murray says. "We ship throughout the world. ... In the North American market, we cover about 85 percent of the natural quartz stone market."

Efficient freight movement plays a key role for the company. "In the Cambria world," he adds, "logistics make up a significant portion of the total cost of our product."

## Q: What advantages do Minnesota-based shippers enjoy versus some of the other states?

**Patrick Murray:** Regional shipping has been helpful from a cost and compliance standpoint, Murray says. "We can do the value-stream mapping, identify challenges, and try to come up with solutions."

**George Schember:** Minnesota offers many advantages, Schember says, though the state does need to do better in attracting a more diverse workforce to the industry.

"It's wonderful doing business in Minnesota," he says. "We have laws. We have labor talent. We have trustworthy partners. We have data and information-exchange systems so that we can actually run our businesses. We have infrastructure that is competitive on a global scale. We're a long way from a lot of markets, but we can get there in a fairly efficient manner."

**Jim Carver:** According to Carver, Minnesota's educated workforce provides the state's biggest advantage, especially for finding transportation, logistics, and supply-chain professionals. "This is a great place to be," he says.

**Bruce Abbe:** The entrepreneurial side of Minnesota is strong, Abbe says. "It goes right to our farmers, and the other companies that serve them. ... It's a strength that you can build on and you should always try to support where you can—and that's outstate as well as in the Twin Cities."

## Q: When choosing to locate a facility in Minnesota, what factors do you take into account, and is transportation one of those factors?

**Jim Carver:** "The labor market is pretty much key to any decision around a facility," Carver says, adding that transportation obviously is also a big factor. "For us, it comes down to where are your customers, where is your supply, where is your product coming from that is going to supply that plant, how is it going to get there, what is the cost of moving it, and what is the cost to produce what you are going to produce there."

"The labor market is pretty much key to any decision around a facility."

— Jim Carver

**George Schember:** Supply-and-demand factors are another consideration, Schember says. For example, "Where's the supply of seed, and where's the demand for the products off of the plants?"

**Patrick Murray:** "Our facility is always going to remain in Le Sueur as the primary location," Murray says, "and we will try to get as close as we can to that facility from a logistics standpoint."

**Bruce Abbe:** The situation is different for the Midwest Shippers Association, Abbe says, because it is a trade association. "We have a very, very strong base of Minnesota-based companies that are grounded where they are at," he adds, "close to their farmers—the people that produce, and those factors very much enter into it."

Companies want to do business in Minnesota, Abbe explains, but some members must go beyond Minnesota borders to remain competitive. The key to global competitiveness, he says, is access to all modes. Scoular, for example, which is based in Minnesota, has global reach. The company recently built a state-of-the-art plant for the food-grain soybean business near Chicago, where it can get better access to containerized shipping.

## Q: Do you deal with any bottleneck points, whether it's freight coming into or going out of the state?

**Patrick Murray:** The rail system is one of the biggest bottlenecks for Cambria, Murray says. "We actually have a main rail line that is 50 yards from our back door, but I cannot even use it," he says. "It's hauling my product all day long, and it just goes right by me, right up to Minneapolis, and then it comes right back down using a van load or some other service."

**George Schember:** "Some bottlenecks are economic and not just physical," Schember says, "and so we've advocated for more balanced and open switching agreements between the railroads that allow us to serve more markets with facilities. ... We could be sitting on a railroad that's doing switching with another railroad right next to us and we can't access the market that second railroad serves because, basically, we have an economic barrier. So I think you've got to think about bottlenecks on multiple levels—physical, economic, policy."

It also is important, Schember says, to pay attention to what other states and nations are doing to eliminate their bottlenecks because it's a competitive industry, regionally and globally. Other states, such as Iowa, and other countries, such as Brazil, are developing their infrastructure and transportation capability.

**Jim Carver:** According to Carver, bottlenecks also exist in the labor market, such as with driver shortages, and in other ways, such as with truck weight restrictions. Nearby states, for example, allow hauling heavier loads of milk. "That's something we haven't done as a state," Carver says, "and it really does drive how we compete with those states in that market. It makes it more difficult."

**Bruce Abbe:** Abbe agrees that the issue of truck weight in Minnesota needs more attention, especially given that other states allow greater weights.

He also says rail service is key to moving commodities, but the railroads have run into their own challenges, trimming away at short lines. "We're not going to build any more railroads around this country," Abbe says. "We don't have the land to go rolling through in any significant way, so we need to find a way to utilize these systems more."

## Q: Is your company going to do anything different to address truck parking?

**Jim Carver:** Land O'Lakes is evaluating its facilities for more parking, Carver says. The company also is exploring the addition of driver facilities to make it easier for drivers.

**George Schember:** Cargill generally approaches truck parking differently. "We focus on scheduling and throughput," Schember says, "and how we can get the system moving faster because we are high-volume."

**Patrick Murray:** Cambria has one facility, and there's a rest area next to the building. "We actually have the first right of purchase for that rest area as we expand," Murray says. "Ultimately, for the blueprint of Cambria and our building, we have designed a full truck-parking center—that's two or three years down the road."

**Bruce Abbe:** The state needs to reconsider rest areas, Abbe says. "Shrinking them isn't the way to go."

There are also parking issues on the river system, he adds. "One of the most crucial issues for the barge system is where we park the dredging material."

### Q: What regulatory factors will pose a big impact on your business as we move into 2017?

**George Schember:** The market adjusts to regulations, Schember says. "But we just have to be realistic about what we are trying to accomplish with the regulation and make sure the regulation delivers the benefit. Too often recently, in the last 10 years, a lot of the regulations have come in and not delivered."

**Jim Carver:** Food safety regulations will have the biggest impact from a time and effort standpoint, Carver says, emphasizing the importance of evaluating whether regulations lead to a societal benefit. "There's been a rush to change the regulation before you see how the last regulation really panned out."

**Bruce Abbe:** Food regulations affect Midwest Shippers Association members, Abbe says, but they will learn how to adapt. "Food safety is important, period—it just is," he says, "so we need to deal with that." Besides food regulations, he points out, customs and border protection, complaints and dispute regulation, and legislative bottlenecks also have an impact on shippers.

# Boon … or Bust?

***What Can We Expect in 2017 in Light of the Election Results?***

Both the federal and state political landscape changed with the election results of 2016. But what do those changes mean for freight transportation? Jason Craig, director of government affairs for C.H. Robinson, and Margaret Donahoe, executive director of the Minnesota Transportation Alliance, offered their insights. C.H. Robinson is one of the world's largest third-party logistics providers. The Minnesota Transportation Alliance is a statewide coalition of organizations including local governments, businesses, labor, transit systems, rail, waterways, and airports.

## Trucking along: The national view

According to Jason Craig, trucking leads the way as the most dominant mode in the United States, and most freight comes in or out of Minnesota by truck. So what does the change in federal administration mean to the trucking industry?

"It is important to understand where leadership of the agencies is in terms of the transition," Craig says. Washington veteran Elaine Chao, who previously served as Secretary of Labor (2001–2009), is the nominee for Secretary of Transportation. "We do know she has a lot of executive experience," Craig adds. "She knows the rule-making process, and she knows how to operate in D.C."

The Federal Motor Carrier Safety Administration (FMCSA), which regulates the trucking industry, faces some uncertainty. Scott Darling has served as interim head for the last 15 months, and it is unclear how long he will continue to serve the new administration.

"The long and short of it is there is going to be a gap in leadership," Craig says, and that gap may slow any big changes. "When your companies and agencies go through a transition and you have a temporary boss, do big things happen? Not really."

There is speculation about what the administration might do with some rules. Many of them are ready for implementation, such as the electronic logging devices (ELD) mandate. "We see that as going forward," Craig says, leaving only a slim possibility for change should there be legislative or court intervention.

The ELD rule takes effect in December 2017. Announced in 2015, with a two-year implementation, the mandate requires all commercial drivers, with some exceptions, to keep an electronic log. "The industry is very concerned about what the impact is going to be," Craig says, including the cost and complexity of implementation and the impact on capacity. "It's hard for people to get a sense for the order of magnitude of this problem."

Approximately 30 to 40 percent of drivers already may be using some type of electronic log, according to a recent survey, and the rules allow for use of cell phones to comply. Low-cost apps are already on the market, Craig says, with one available for $175 plus $30 a month.

Other yet-to-be-implemented rules, such as those that relate to compliance, safety, accountability (CSA) reform, speed limiters, and overtime, may receive new attention.

Infrastructure investment also is important to trucking and will be a topic for discussion at the federal level. "This was a big campaign promise," Craig says. "Is this really going to happen? Because you may or may not remember that we passed a transportation bill, a five-year transportation bill that was a big deal in December of 2015. It provided almost $225 billion dollars over five years—$43 to $47 billion per year, depending on the year. They had a tough time paying for it. So the question is, what are they going to do next?"

Since the transportation bill was authorized for a full five years, Congress doesn't have to do anything on the infrastructure, Craig points out, and there isn't agreement on how to pay for it. "In terms of infrastructure funding at the federal level, the only place I see it happening is tied to a corporate tax reform bill and that has a lot of momentum," he says. "But I have some healthy skepticism about an infrastructure bill at the federal level."

## Pushing forward with momentum: The state view

The Minnesota Legislature, close to approving increased infrastructure investments last year, faces a session that likely will bring continued discussions about those investments.

Freight movement depends heavily on infrastructure, Margaret Donahoe says. "We can't just increase the weight limits and expect the infrastructure can handle that." For example, many bridges in Minnesota were not designed for heavier weights.

But a united front can make a difference. "If we can get everybody working together to really push legislators to increase the investments to improve the strength of the roads and bridges, we could have a real win-win for everybody," she says.

Though Minnesota has had some big construction years, that is not going to last without some action at the state and federal levels to increase the investments, Donahoe says. And while the highway trust fund will be increasing, the purchasing power of those dollars will decrease significantly when adjusted for inflation.

**U.S. Trucking**

**206,667**
**for-hire motor carriers**

**1,003,939**
**truck tractors**

**89%**
**with a carrier fleet size of 1–5 trucks**

**0.1%**
**with a carrier fleet size of 400 or more trucks**

*Note: LTL, courier, and private fleets not included.*

(Source: C.H. Robinson)

**Minnesota State Highway Investment Plan (MnSHIP)**

**2014–2033**
**Revenue available:**
$18 billion
**Total need:**
$30 billion

**2018–2037**
**Revenue available:**
$21 billion
**Total need:**
$39 billion

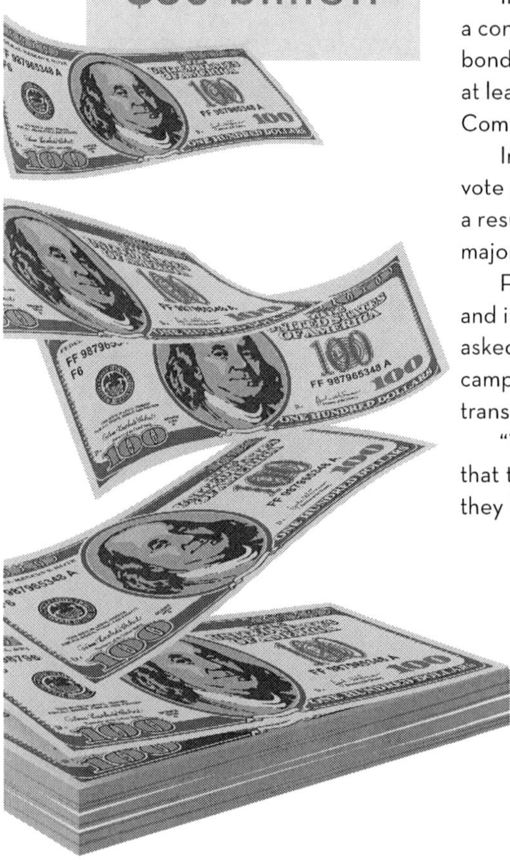

The state is quickly falling behind in its infrastructure investment. The Minnesota Department of Transportation 20-year plan projects revenue and needs for the system. "When they came up with the plan for 2014 to 2033, the funding gap was about $12 billion over 20 years," she says. "The latest plan shows that, yes, the resources are increasing, but so is the need. And so now the gap has increased to $18 billion over 20 years."

What's more, much of the investment is focused on maintaining and preserving the existing system, with very little targeted for adding capacity or expansion of the system.

On the positive side, Donahoe says, the state did establish the Corridors of Commerce program to identify the corridors with the most impact on moving freight throughout Minnesota. Thanks to some trunk highway bonding, the state prepared some projects for construction, but funding has yet to be approved for the actual work.

In the last few years, both the governor and legislative leaders expressed support for transportation as a priority. "There were agreements between both parties—at least on the highway side—that the shortfall is $600 million a year and there should be a way to find that $600 million a year on an ongoing basis," she says. "But unfortunately there was disagreement about how to find that $600 million a year or how to fund transit, especially in the metropolitan area."

In 2015, separate Minnesota House and Senate bills never emerged from a conference committee. The Legislature then proposed a substantial capital bonding bill as a way to increase funding for transportation infrastructure—at least on a one-time basis—including some funding for the Corridors of Commerce. "Unfortunately," she says, "the bonding bill didn't become law."

In 2017, the GOP has a majority in the House and Senate—but only a one-vote majority in the Senate—and the Democratic governor has veto power. As a result, Donahoe says, it likely will take bipartisan cooperation to get anything major accomplished.

For this year's legislators, health care and tax relief, as well as transportation and infrastructure, are big issues. "There were a lot of candidates that were asked about transportation, who talked about transportation during the campaign season, who talked about the need for a long-term, comprehensive transportation funding package.

"We need to keep reminding legislators that they didn't get the job done that they promised to get done in the last biennium," Donahoe concludes, "and they have a really good opportunity in 2017."

## Global Supply Chains Begin at the State/Regional Level

University of Minnesota
*Center for Transportation Studies*
*19th Annual Freight & Logistics Symposium*
*IHS Economics*
*Transportation & Trade*

**By**
**Charles W. (Chuck) Clowdis**
**Managing Director – IHS Economics**

# Data IS Available....

- Federal Government
- State Agencies
- MPOs
- Commercial sources....
- Custom-Tailored Research
- Thru blood, sweat & tears....

# IHS Transearch© Overview

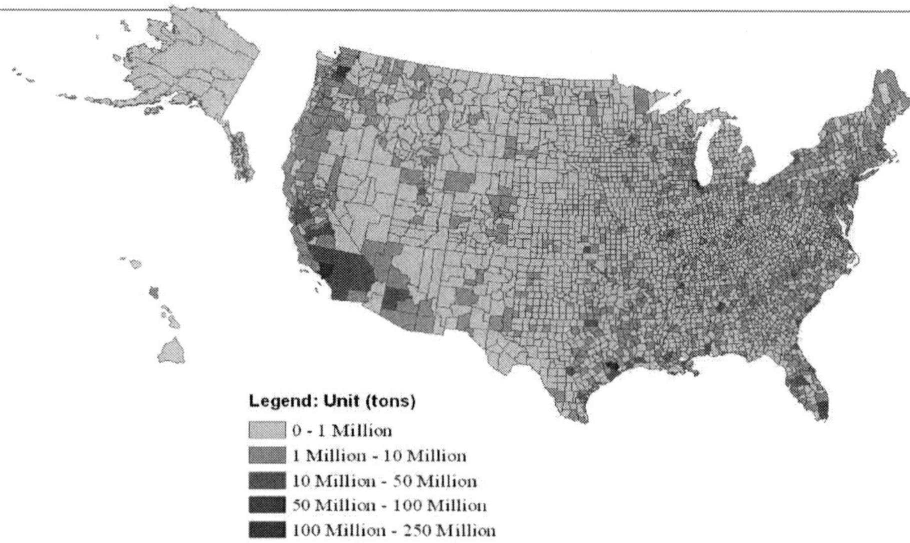




# What is IHS Transearch©?

- Unified multi-modal goods movement database
- Standard source of US freight flow information
- Shows commodity by volume, mode and lane
- Integrates large scale proprietary truck flow sample with private and public source data
- Maps flow data to GIS networks
- Flow detail down to county level
- Supply Chain design, support, implementation options
- Consistent construction for current and forecast years
  - Covers both short and long term

## IHS Transearch© Features

| Provision Of Data Elements | |
|---|---|
| **In Product** | **On Custom Basis** |
| All truck shipments of Manufactured Goods:<br>• Primary & Secondary<br>• TL, LTL, Private<br><br>All truck shipments of Agricultural Products, Coal, Ores, & Minerals<br><br>Empty trucks<br><br>All rail, water, air shipments<br>• Plus air & intermodal truck dray<br><br>All US/Canada; US/Mexico<br><br>Inland Movement of Overseas Shipments (Imports & Exports) | Truck shipments of:<br>• Fish & Forest products<br>• Waste<br>• Mail, Express, Government<br><br>Domestic water drayage<br><br>Sub-county detail<br><br>Local data integration:<br>***NEW Supply Chain OPTION, Fall 2016*** |



## How Can IHS Transearch support a SC design?

- Raw materials vendor/source locations
- Modes and modal options
- Supplant with IHS Freight Finder® to locate exact Vendors & Locations
- Add IHS World Trade Service® data for Import/Export support
- Routing tool to determine best routes for all modes

## Is "no data" riskier than erroneous data?

"Without IHS Markit Transearch © proven, reliable, verified, freight flow data; you're simply a group of planners with your own opinions".

 

## Supply Chain Optimization Options

 

# Identify & Quantify Location Options

- By County
- By BEA
- By City, Metro Area, ZIP
- By Transportation Analysis Zones (TAZ), Industrial Parks, etc.
- By proximity to :
  - Major highways, intersections, on/off ramps, access roads, transloading facilities
  - Major Rail lines
  - Intermodal Ramps
  - Airports
  - Inland Ports
  - Intermodal yards
  - Free-trade Zones
  - Rivers, barge terminals
  - Suppliers and Vendors
  - Workforce demographics

 9

# Location Parameters (continued)

Transportation Rate* and Availability Benchmarks

- Motor Carrier – by type of equipment, service, etc.
- Rail – by equipment type, service, etc.
- Drayage – by type to/from transloading facilities, ramps, airports, Customers & Vendors
- Barge – to/from river terminals via truck, rail
- Warehousing availability, pricing levels, etc.
- Vendor identification by industry needs
- Labor Demographics
- Work-force availability
- Work-force Training availability

***Rate information derived from proprietary sources; for example, truck rates are sourced from paid freight bill data. No contractual covenants are breached between Shippers or service providers. Other rates/costs are derived from publically available data.**

 10

## Transport Mode Options

List of all commercial transport modes serving the specified target area:

-Contact information for service providers

-Assistance with Requests for Indication of interest (RFIs) and Requests for Service/Price Proposals (RFPs)

-Management support of RFP responses (review, selection, implementation) also available





## Simplified, Basic Supply Chain Design

### Basic Supply Chain

From raw materials sources to
Factory to DC and beyond......

Illustration by Chris Gash

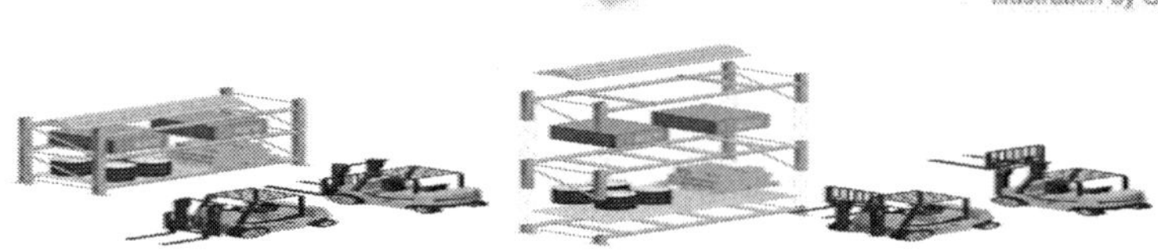

# IHS Transearch© Source Data

**Primary domestic traffic flow sources**

- Annual Motor Carrier Data Exchange (Proprietary Shipment Data)
- Annual STB Railroad Waybill Sample
- Annual Rail Carrier Data Exchange (Proprietary Shipment Data)
- Annual Corps of Engineers Waterborne Commerce Statistics
- Annual US DOT Airline & Airport Statistics
- US Customs Import/Export Trade Statistics
- Annual Department of Energy Coal Movement Statistics
- Bureau of Transportation Statistics - Commodity Flow Survey

**Primary production and shipments sources**

- Census County Business Patterns
- US Geological Survey Commodity Reports
- Annual Department of Agriculture Crop & Livestock Data
- Annual Railroad Freight Commodity Statistics
- IHS Industrial Sector Production & Business Transactions
- IHS Automotive Plant & Production Forecasts
- IHS Business Market Insights (Wage, Population & Economic Data)
- Inter-Industry Trade Patterns (Input/Output Table)

There are over 100 individual data sources used in the construction of Transearch



# IHS Transearch© Public Source Data

| | | |
|---|---|---|
| US GEOLOGICAL SURVEY MINERAL INDUSTRY REPORTS | DEPT OF THE INTERIOR, U.S. GEOLOGICAL SURVEY | QUESTIONNAIRES SENT TO PRODUCERS, PLANTS, DISTRIBUTORS, IMPORT TERMINALS |
| US DEPT OF AGRICULTURE CENSUS & ANNUAL SURVEYS | US & STATE DEPT.S OF AGRICULTURE | CROP & LIVESTOCK PRODUCTION FROM FARMS & RANCHES |
| COUNTY POPULATION DATA | DEPT OF COMMERCE, CENSUS BUREAU | CENSUS OF POPULATION COLLECTED EVERY 10 YEARS AND ESTIMATED ANNUALLY |
| INTER-INDUSTRY TRADE PATTERNS (INPUT/OUTPUT TABLE) | DEPT OF COMMERCE, BUREAU OF ECONOMIC ANALYSIS | INPUT-OUTPUT (IO) ACCOUNTS, BASED ON ECONOMIC CENSUS DATA |
| PRIVATE PORT DIRECTORIES | VARIOUS BARGE LINES, PORTS | PROPRIETARY INFORMATION ABOUT WATERWAY FACILITIES, COMPILED BY SURVEY |
| DEPT. OF ENERGY COAL MOVEMENT STATISTICS | DEPT. OF ENERGY | MONTHLY, QUARTERLY, ANNUAL SURVEYS OF ELECTRIC UTILITIES; EXIM CUSTOMS DATA |
| BTS COMMODITY FLOW SURVEY | BUREAU OF TRANSPORTATION STATISTICS | COMPULSORY, STRATIFIED SURVEY: MANUFACTURERS, WHOLESALERS, SOME RESOURCE PRODUCERS |
| STATISTICS INTERNATIONAL TRADE DATA | STATISTICS INTERNATIONAL TRADE DIVISION | COLLECTED FROM CUSTOMS DECLARATIONS, IMPORT ENTRY FORMS |
| BTS TRANS-BORDER STATISTICS | BUREAU OF TRANSPORTATION STATISTICS | SHIPPER EXPORT DECLARATIONS, IMPORT ENTRY FORMS |
| COE WATERBORNE COMMERCE STATISTICS - STATE TO STATE SERIES | ARMY CORPS OF ENGINEERS | COMPULSORY SURVEY OF OWNERS, AGENTS, MASTERS, CLERKS OF VESSELS ON NAVIGABLE WATERS |
| COE WATERBORNE COMMERCE STATISTICS - PORT SERIES | ARMY CORPS OF ENGINEERS | COMPULSORY SURVEY OF OWNERS, AGENTS, MASTERS, CLERKS OF VESSELS ON NAVIGABLE WATERS |
| BTS T-100 DOMESTIC & INTERNATIONAL TRAFFIC DATA | BTS, OFFICE OF AIRLINE INFORMATION | REPORTS OF U.S. AND FOREIGN CARRIERS OF FREIGHT AND/OR MAIL |
| BTS FORM 41 T-3 ENPLANEMENT STATISTICS | BTS, OFFICE OF AIRLINE INFORMATION | REPORTS OF CERTIFICATED AIR CARRIERS BY |
| FAA 5010 AIRPORT DATABASE | DEPT OF TRANSPORTATION, FEDERAL AVIATION ADMIN | COMPILATON OF CERTIFICATED LANDING FACILITIES |
| STB RAILROAD WAYBILL SAMPLE | SURFACE TRANSPORTATION BOARD | STRATIFIED, COMPULSORY WAYBILL SAMPLE FOR RAILROADS TERMINATING 4000+ LOADS/ YEAR |

# IHS Transearch©
# Public Resources (examples)

| Source | Q1 | Q2 | Q3 | Q4 |
|---|---|---|---|---|
| USGS Mineral Reports | | X | | |
| USDA Agriculture Surveys | X | X | | |
| BTS/FAA Air Activity | X | X | | |
| DOE Coal Movements | | X | | |
| BTS Trans-Border Data | X | X | | |
| STB Waybill | | | X | |
| COE Waterborne Commerce Data | | One Year Lag | | |



15

# TRANSEARCH Processing Overview

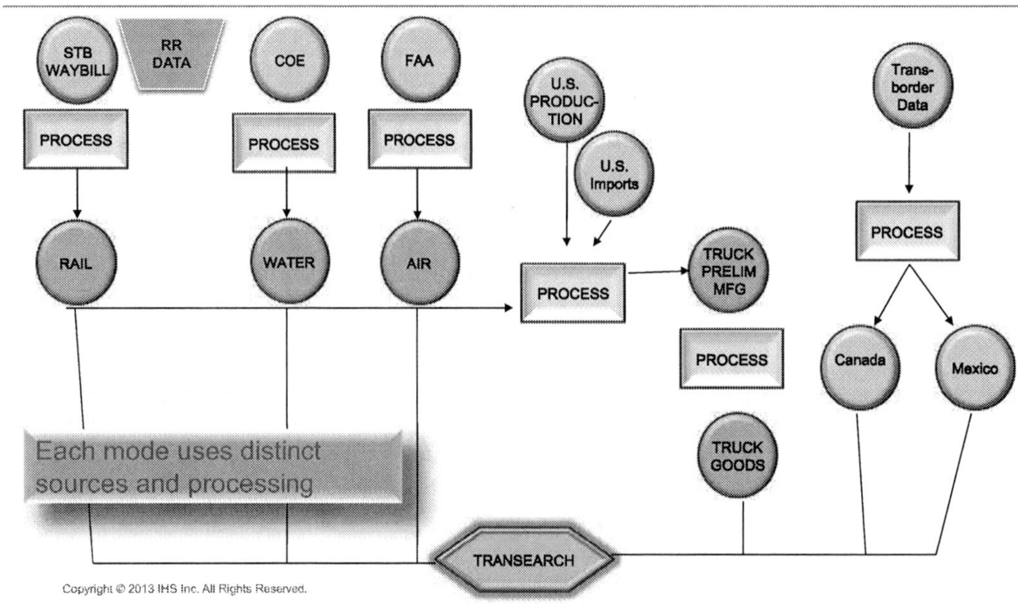

## IHS Transearch© Forecasts

- Long-term annual forecast of all base year IHS Transearch dimensions to 2045.
- Regional input/output modeling approach with industry establishment output and purchases and consumer purchases from the IHS Business Demographics Model at county level.
- Captures geographic shifts in production, population and sourcing.
- County growth constrained to National-level



## Transearch Forecast Process

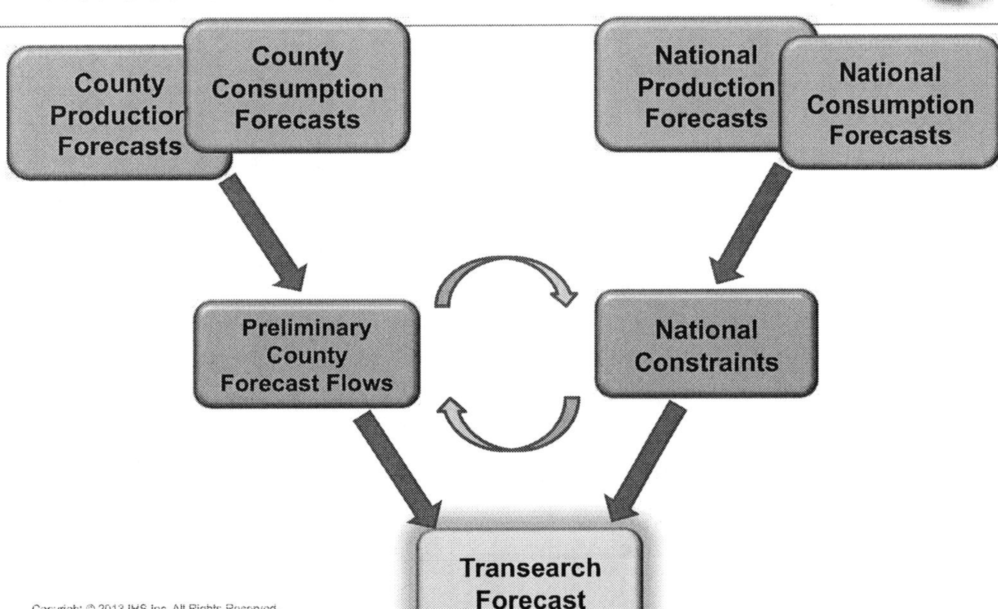

## Transearch Generations (The Reebie years)

- Before 1980 Traffic Year:
  - Custom Project-Specific Data Sets
- 1980's:
  - State/BEA Level US Data
  - Canada Province Cross-Border Flows
- 1990's:
  - Motor Carrier Data Exchange
  - NAFTA Gateway & Mexico Cross-Border Flows by State
  - Distribution Center/Wholesale Traffic
  - County-level market coverage
  - Freight Analysis Framework
  - Ores/Minerals
  - Drayage Traffic



## Transearch Generations

- 2000's:
  - Expanded coverage of Agriculture
  - Rail Data Exchange
  - Global Insight Production Drivers utilized
  - Empty Truck Traffic
  - 2nd Generation Agriculture & Minerals-enhanced detailed coverage
  - Global Insight Integration: Business Sectors & Transactions data
  - Import/Export Identification on final records
  - Full Industry I/O Consumption Quantification
  - IHS Source Data Integration & QC Process
  - Integration of forecasts from other units within IHS
  - Polk Automotive Data Integration

## NAFTA Detail

- Canada/Mexico to and from United States movements only
  - No Canada-to-Mexico or Mexico-to-Canada data in Transearch
- Geographical Detail
  - Canada: Province & Canadian Metropolitan Area (CMA) where applicable
  - Mexico: Mexican state only
  - US: County of ultimate origin/termination and border crossing county
- Sources
  - Primary sources are US Customs data and data from the Bureau of Transportation Statistics
    - Only shipments reported by customs are included in Transearch
  - Supplemental sources are Transport Canada data for Canada and the Mexico Economic Census (ENEGI) and annual port statistics (SCT)



## Sample Move: Rail Intermodal & Drayage

- Transearch move consists of three parts:
  - Intra-BEA truck move in Los Angeles BEA (160)
    - STCC: 5021 (Rail Drayage to Ramp)
    - Tradetype: Import
  - Rail Intermodal move from Los Angeles to Illinois portion of Chicago BEA (64)
    - STCC: 4611 (Freight all kinds)
    - Tradetype: Import
  - Intra-BEA truck move in Chicago
    - STCC: 5022 (Rail Drayage from Ramp)
    - Tradetype: Import

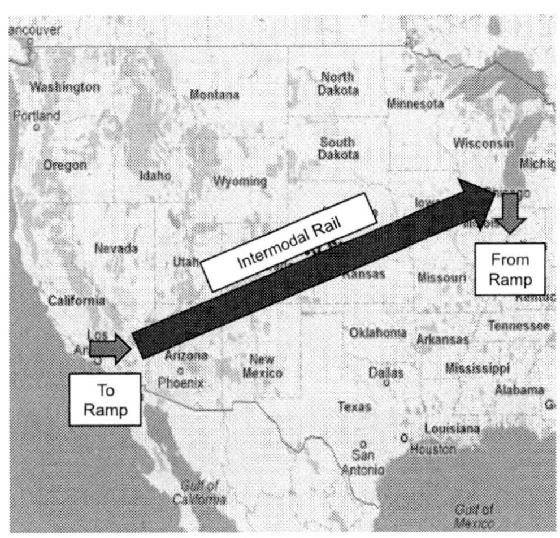

# Constructing IHS Transearch (sample model)

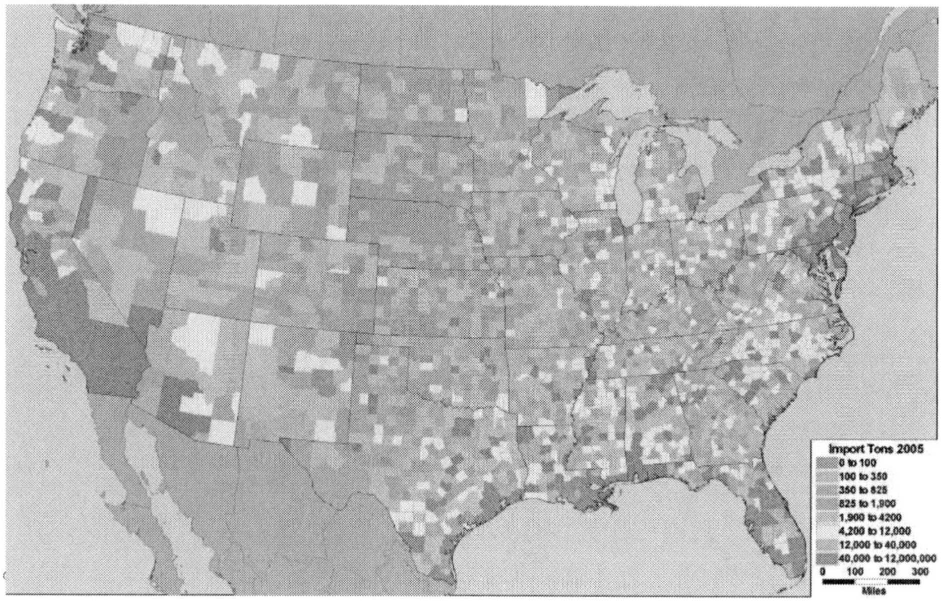


# Business Market Insights® (BMI®) Coverage

- **IHS Global Insights Business Market Insights® database consists of the following concepts:**
  - Number of Establishments
  - Employment
  - Sales (Nominal & Real Dollars)
- **Employment, establishment and sales are developed for all 6 digit NAICS code industries.**
- **Historical county level data available from 1990 to present for employment and establishments and from 1997 to present for sales**
- **Forecast extends 25 years into the future**
- **Data available for all of the nations 3,144 counties**

## IHS Transearch Quality Control Process- The IHS Commitment to "Customer Delight"

**Annual Development Process**

- Continuous Client Feedback
- Comparisons with Historic Transearch Patterns
- Comparisons with other Data Sources: CFS, FAF, Industry Data
- Verification of Select Facilities via Aerial Imagery

**Order Processing**

- Pricing Proposal Creates Order Specification Sheet
- Incorporate Client-specific data
- Comparison of Results with Data Summaries & Previous Orders
- Results Review by Additional Staff



## Future of IHS Markit Transearch©

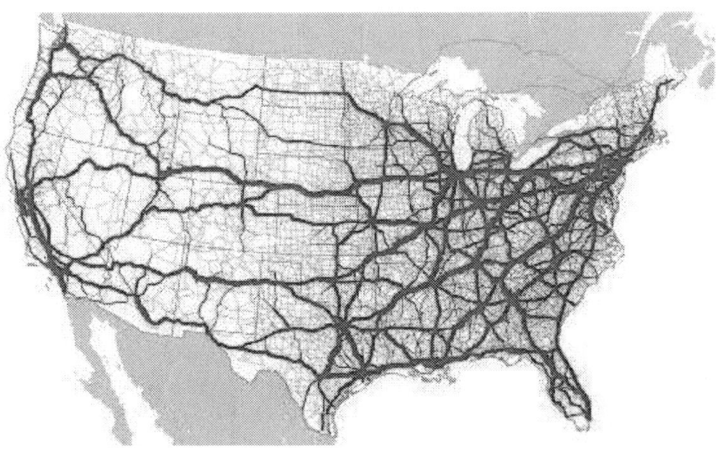

## The "Heart of America" (sample from Transearch® 2014)

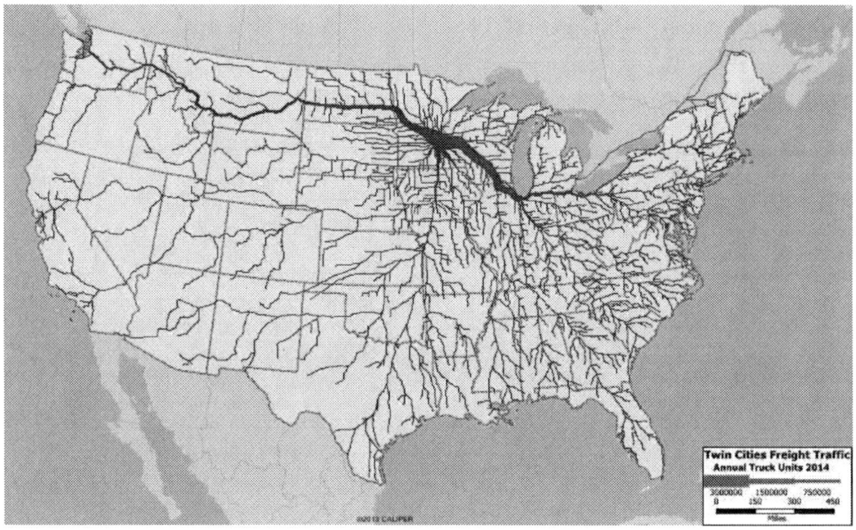






## IHS Markit **Point of View:** Supply Chain Freight Data is ***NOT*** the same as Freight Flow Data

| Freight Flow Data | Supply Chain Data |
|---|---|
| Where & How a Shipment Moves | What Industry Segments Generates & Consumes the Shipments |
| Market-to-Market Activity | Follow All Steps through an Industry |
| Extensive use of Supply/ Demand Models | Significant input from specific Industry Experts |

***Supply chain freight planning requires industry data and expertise…………..***

***IHS CAN HELP!***

## Auto Supply Chain

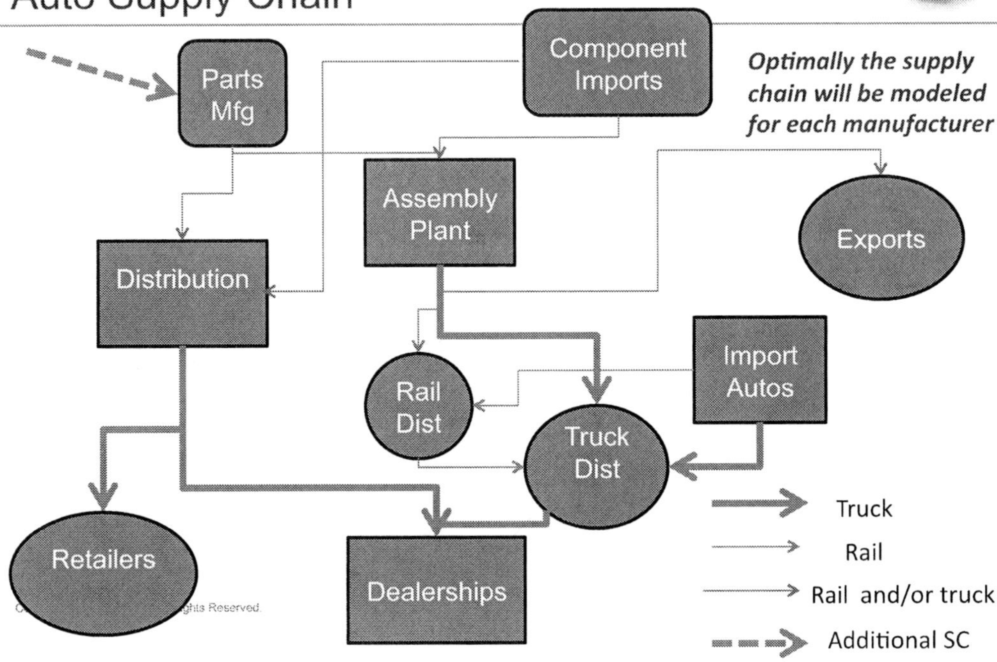


## Going Deeper: Major Chemical Plants in DE

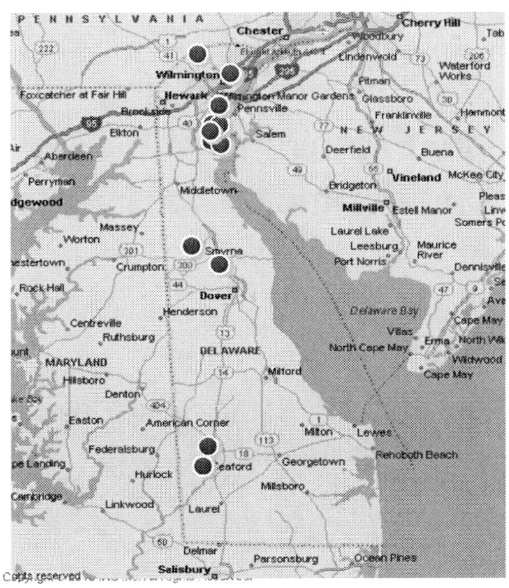


| Company | State | City |
|---|---|---|
| BASF Corp. | DE | Seaford |
| Chesapeake Green Fuels | DE | Clayton |
| Croda Chemicals | DE | New Castle |
| Delaware City Ref. | DE | Delaware City |
| DuPont | DE | Edge Moor |
| DuPont Chem. | DE | Delaware City |
| FPC USA | DE | Delaware City |
| Invista | DE | Seaford |
| Kuehne | DE | Delaware City |
| NVF | DE | Yorklyn |
| Reichhold | DE | Cheswold |
| Standard Chlor. | DE | Delaware City |

## Supply-Chain Data Improves Freight Flow Development

| Methodology | Existing | Supply-Chain Detail | |
|---|---|---|---|
| Production Commodity Detail | 4 digit | 5 digit | |
| Product Example | 2812-Potassium or Sodium Compounds | 28128-Chlorine | |
| Market Detail | County | 53 Production Plants (US); 1 in DE | |
| Consumption Allocation | I/O Analysis | 11 Detailed Uses - | |
| Consumer Detail | County | Chlorinated Intermediates (5%) | 8 Plants |
| | | Epichlorohydrin (6%) | 2 Plants |
| | | Inorganics (6%) | 11 Plants |
| | | MDI (5%) | 4 Plants |
| | | Others (21%) | 66 Plants |
| | | Polycarbonates (2%) | 4 Plants |
| | | Propylene Oxide (8%) | 2 Plants |
| | | Pulp & Paper (1%) | 11 Plants |
| | | TDI (3%) | 2 Plants |
| | | Vinyls (39%) | 23 Plants |
| | | Water Treatment (6%) | up to 15,000 wholesale/ distributors |



## Using IHS Transearch©

- Modal, capacity, and infrastructure planning
- State and regional Freight Planning
- Assessments of the economic impact of domestic & foreign trade
- Freight corridor studies
- Supply Chain Optimization
- Alternative transportation and diversion studies
- Network modeling such as rail, highway, and water

## Transearch Redesign Underway – change dates

- In 2012, IHS began a top-to-bottom review of all of the processes involved in assembling Transearch
  - Goals: More accurate, more streamlined production, more consistent with other IHS data, more reflective of clients' needs.
- Rolled out in TS2011
  - Integration with CMAI chemical data – netting of same & near plant chemical flows; improved Ethanol processing
  - Separate processing for West Coast petroleum products
- **Included in** TS2012
  - Redesign of Gravity Model
  - Crop specific agriculture processing
  - More integration with Downstream team (former Purvin & Gertz)
- Help us make Transearch fit your needs!




## IHS Transearch© Usage: a few Case Studies

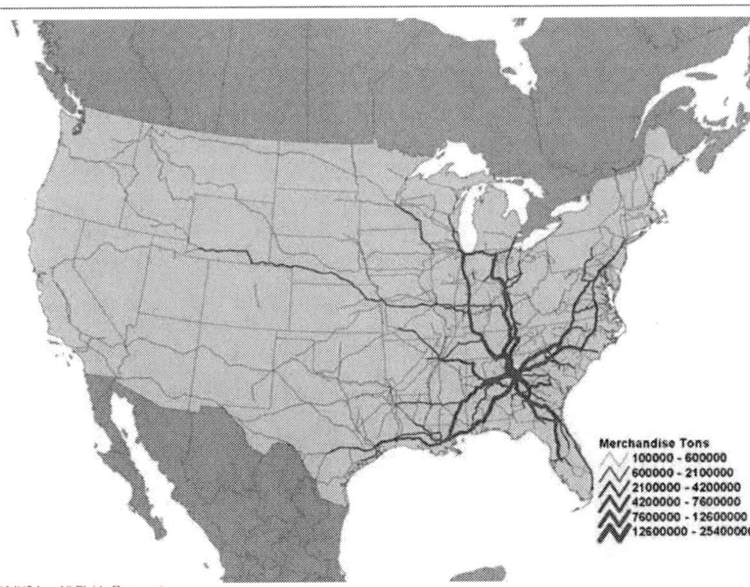

## Rail Capacity: Meeting Future Needs

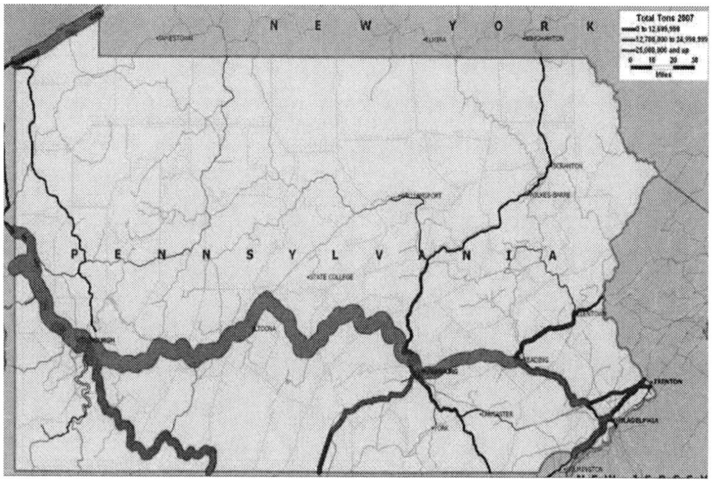


**Need: What freight moves on Pennsylvania's rail network? What routes and counties will grow in the next 50 Years?**

**Solution: Use TRANSEARCH and Federal rail data to identify and forecast Pennsylvania rail flows. Combine with rail carrier capital planning projects.**



## Market Competitiveness

**Need:** A Chamber of Commerce needed an assessment and analysis of its region's current and projected infrastructure capacity and supply chain efficiencies; and a infrastructure development program.

**Solution***:* A Supply Chain Optimization & Capacity Assessment of multimodal freight linkages, infrastructure and activity centers was conducted integrating surveys, interviews, *Transearch*, and *Freight Finder®*. *IHS* Transearch© forecasts of freight flows demonstrated future needs, and Freight Finder® identified facilities, their supply chains, and aided in prioritizing structure needs.

**Infrastructure Analysis to Support Key Sectors and Companies**

**Public Private Partnership for Transportation Improvements**

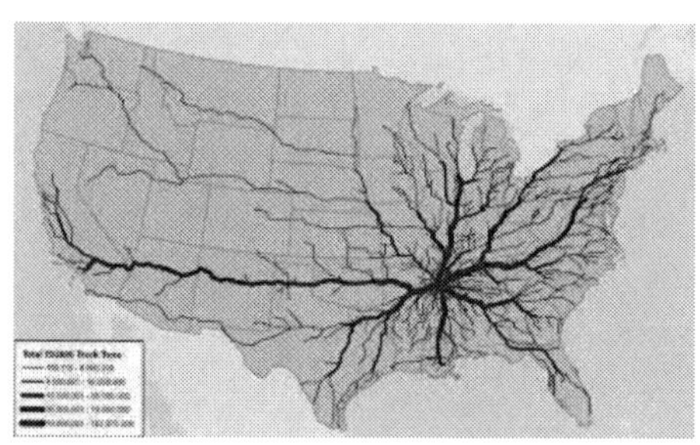

## Inland Intermodal Port Market Assessment

**Need:** An inland port authority had the opportunity to build a multimodal facility with funding and land contributed by a Class I, if they proved its viability and gained additional funds from the State. The client needed a fact-based and credible business opportunity & infrastructure analysis.

**Solution:** ***Transearch*** determined regional freight volumes by mode & route, and quantified the potential to divert freight to the site. ***Freight Finder*** identified key industries to attract, and the ***IHS Transportation Economic Impact Model*** evaluated the economic impact of the facility on the State.

**If, Where and How to Build an Inland Port**

## Moving Forward: Data Exchange Program

Historic & Current Partners

- Motor Carriers 1990
- Class I Railroads 1995
- 3rd Pty. Providers, Load Matching Services 1999

***More Potential New Partners***

- Local/dray Trucking Companies added 2013
- GPS Satellite, Barge, air cargo Data added 2015
- Local Surveys & Studies on- added 2015

## IHS Transearch©: Moving Forward

**From This (Commodity Flow):**

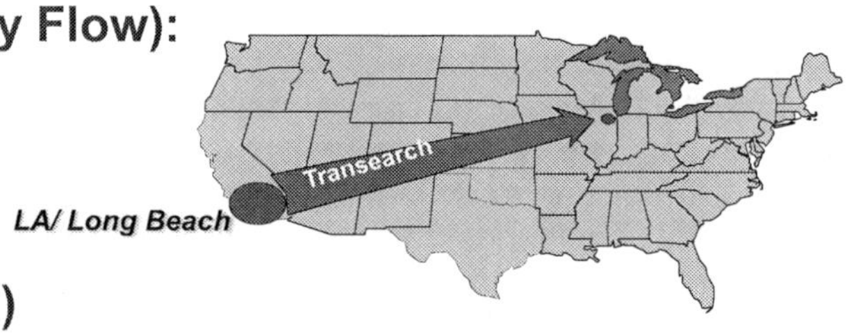


**To This (Supply Chain)**

**adding Fall 2016!**

*See earlier slides for initial details*

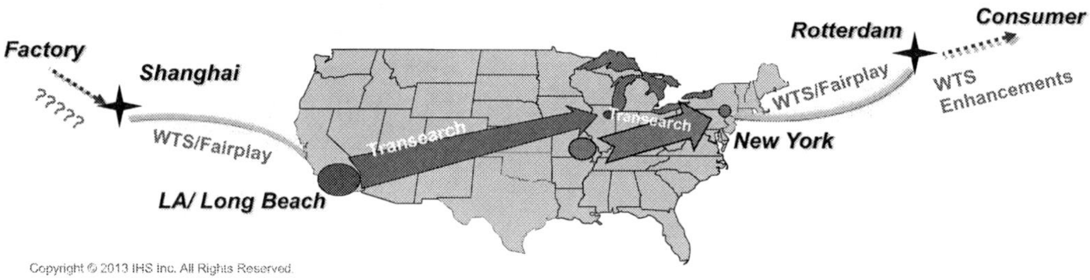




## **IHS Point of View:** Supply Chain Freight Data is ***NOT*** the same as Freight Flow Data

| Freight Flow Data | Supply Chain Data |
|---|---|
| Where & How a Shipment Moves | What Industry Segments Generates & Consumes the Shipments |
| Market-to-Market Activity | Follow All Steps through an Industry |
| Extensive use of Supply/ Demand Models | Significant input from specific Industry Experts |

***Supply chain/freight flow planning requires BOTH industry data and expertise………….. IHS CAN HELP!***

## IHS has a unique set of expertise

Industry-expert business groups
- Agriculture
- Automotive
- Chemicals
- Construction
- Consumer & Retail
- Energy
- Metals & Mining
- JOC/PIERS®, Fairplay

Information Products
- Automotive supplier database
- Chemical databases
  - Supply & Demand
  - Capacity

World Trade Service



## Questions?

# THANK YOU!

**For more information and/or details on licensing of IHS Transearch© & Supply Chain Optimization Studies, contact:**

Charles W. (Chuck) Clowdis
**Managing Director – IHS Economics, Transport**
**+ 1 781.301.9020**
charles.clowdis@ihsmarkit.com

## National Freight Issues
## 2017 Outlook

Jason Craig
Director of Government Affairs

»Accelerate Your Advantage® C.H. ROBINSON

» Current Leadership

Secretary of Transportation
Anthony Foxx

FMCSA Administrator
T. F. Scott Darling, III

2

»Accelerate Your Advantage® C.H. ROBINSON

## » Future Leadership

Secretary of Transportation
Elaine Chao

FMCSA Administrator
TBA

3

» Accelerate Your Advantage  C.H. ROBINSON

» Infrastructure Investment

## 2015 FAST Act (Dec 2015)

- 5 year bill through Fall of 2020
- Provides $43-$47 Billion per year
  - $225 Billion in total
- Fully "paid for" including:
  - Indexing customs fees to inflation
  - Selling of Strategic Petroleum Reserve
  - Allow IRS to hire private tax collectors

## 2017 Infrastructure Investment

- Forces against a new bill
  - No required deadline
  - No agreement on how to pay for
  - States stepping up
- Momentum for a new bill
  - Core campaign promise
  - Construction jobs
- Options for "pay fors"
  - Repatriation of foreign earnings
    - Requires tax reform
  - Oil royalties
  - Private money (PPP)
  - Series of Smaller Bills
  - Barrel Tax or Gas Tax

»Accelerate Your Advantage® C.H. ROBINSON

» Regulatory Hot Topics

## 2017-18 Impact

- Coercion Rule
- ELD's
- HOS fix
- Sanitary Transportation of Food
- Overtime Rules*
- Drug and Alcohol Clearing House
- Other Issues:
  - Public Liability Insurance Minimums*
  - CA break rules (F4A issue)
  - ACE Customs Rollout
  - US DOT # Updates
  - TPP*

## 2018-19 and beyond impact

- Speed Limiters*
- CSA Reform*
  - SFD rulemaking is stalled until correlation study is completed
- Other Issues:
  - Detention/Driver Pay*
  - Emerging Technology (Autonomous Trucks, Drones, etc)

»Accelerate Your Advantage® C.H. ROBINSON

## » ELD's

### Timeline:

- Implementation is required by December 2017
- OOIDA Lawsuit was denied in Nov 2016

### Major Issue(s):

- Mandates ELD's for all motor carriers
  - Impact of ELD's and Driver Harassment
- Cost for compliance
  - Smartphone apps can be compliant
  - Costs as low as $175 for hardware and $30/month
- Technological complexity of details
- Impact on capacity?

### Impact to Shippers:

- Will fear or fundamentals impact the market?
- Check back in March 2017 for an update

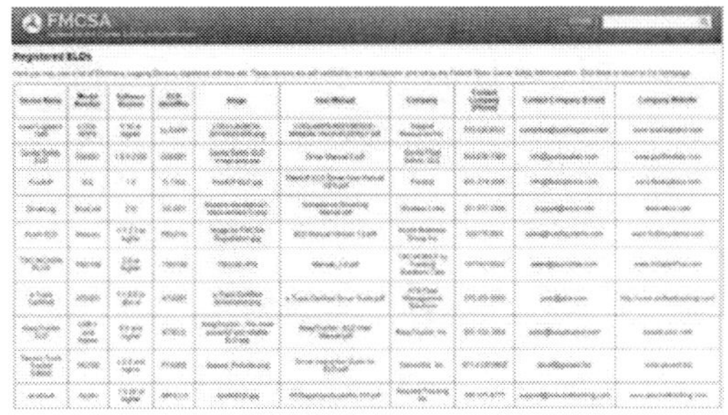


»Accelerate Your Advantage C.H. ROBINSON

## » ELD's

### How many HOS violations are there?

- 2015 CVSA roadcheck data placed 2.88% of all inspected drivers out of service for HOS and False Log violations
  - For comparison, 12.4% of vehicles were placed out of service during the surprise brake check in May 2016
- 2016 FMCSA Pocket Guide to Truck Statistics cites 3.3 million roadside inspections conducted in 2015 and a maximum of 13.7% of violations related to logbook violations.
  - Only 23,652 violations were written for driving more than 11 hours (0.7% of all stops)

2-10 Most Frequent Driver Violations in Roadside Inspections, 2015

| Violation Code | Category | Violation Description | Number of Violations |
|---|---|---|---|
| 395.8 | No Log/Log Not Current | Log Violation (General/Form and Manner) | [illegible] |
| [illegible] | All Other Driver Violations | Non-English Speaking Driver | [illegible] |
| [illegible] | No Log/Log Not Current | Driver's Record of Duty Status Not Current | 72,045 |
| [illegible] | Hours of Service | Driving Beyond 8-Hour Limit Since the End of the Last Off-duty or Sleeper Period of At Least 30 Minutes | [illegible] |
| [illegible] | Traffic Enforcement | State/Local Laws - Speeding 6-10 Miles Per Hour Over the Speed Limit | [illegible] |
| [illegible] | Seat Belt | Failing to Use Seat Belt While Operating Commercial Motor Vehicle (CMV) | [illegible] |
| [illegible] | Traffic Enforcement | Failure to Obey Traffic Control Device | [illegible] |
| [illegible] | Hours of Service | Driving Beyond 14-Hour Duty Period (Property-carrying Vehicle) | [illegible] |
| [illegible] | Medical Certificate | Operating a Property-carrying Vehicle without Possessing a Valid Medical Certificate | [illegible] |
| [illegible] | No Log/Log Not Current | False Report of Driver's Record of Duty Status | [illegible] |
| [illegible] | Medical Certificate | Driver Not in Possession of Medical Certificate | [illegible] |
| [illegible] | No Log/Log Not Current | Driver Failing to Retain Previous 7 Days' Logs | [illegible] |
| [illegible] | No Log/Log Not Current | No Driver's Record of Duty Status | [illegible] |
| [illegible] | Hours of Service | Driving Beyond 11-Hour Driving Limit in a 14-Hour Period (Property-carrying Vehicle) | 23,652 |
| [illegible] | Traffic Enforcement | State/Local Laws - Speeding 11-14 Miles Per Hour Over the Speed Limit | [illegible] |
| [illegible] | All Other Driver Violations | Operating a CMV without a Commercial Driver's License (CDL) | [illegible] |
| [illegible] | Traffic Enforcement | Lane Restriction Violation | [illegible] |
| [illegible] | All Other Driver Violations | Using a Handheld Mobile Telephone While Operating a CMV | [illegible] |
| [illegible] | Traffic Enforcement | State/Local Laws - Speeding 15 or More Miles Per Hour Over the Speed Limit | [illegible] |
| [illegible] | Medical Certificate | Expired Medical Examiner's Certificate | [illegible] |

Notes: Total number of driver inspections in 2015: [illegible]. Total number of driver violations in 2015: [illegible]. Total number of driver out-of-service (OOS) violations in 2015: [illegible]. Only the top 20 driver violations (based on frequency of occurrence) are listed in this table.
Data Source: FMCSA, Motor Carrier Management Information System (MCMIS), data snapshot as of January 29, 2016.

»Accelerate Your Advantage C.H. ROBINSON

Thank You

C.H. ROBINSON

# Minnesota Transportation Alliance Election Impact December 2, 2016

THE TRANSPORTATION ALLIANCE

Highways • Transit • Rail • Waterways • Air

# Construction Outlook

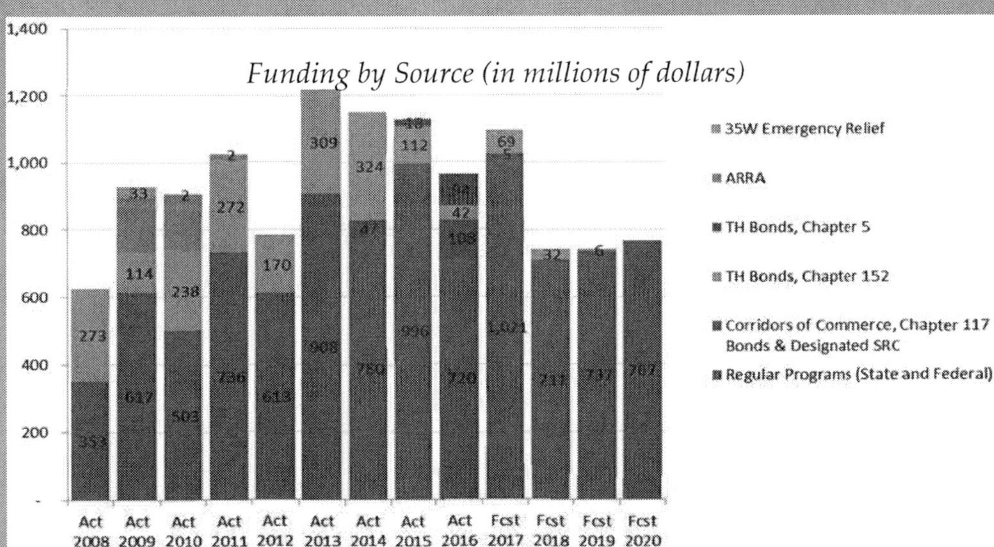

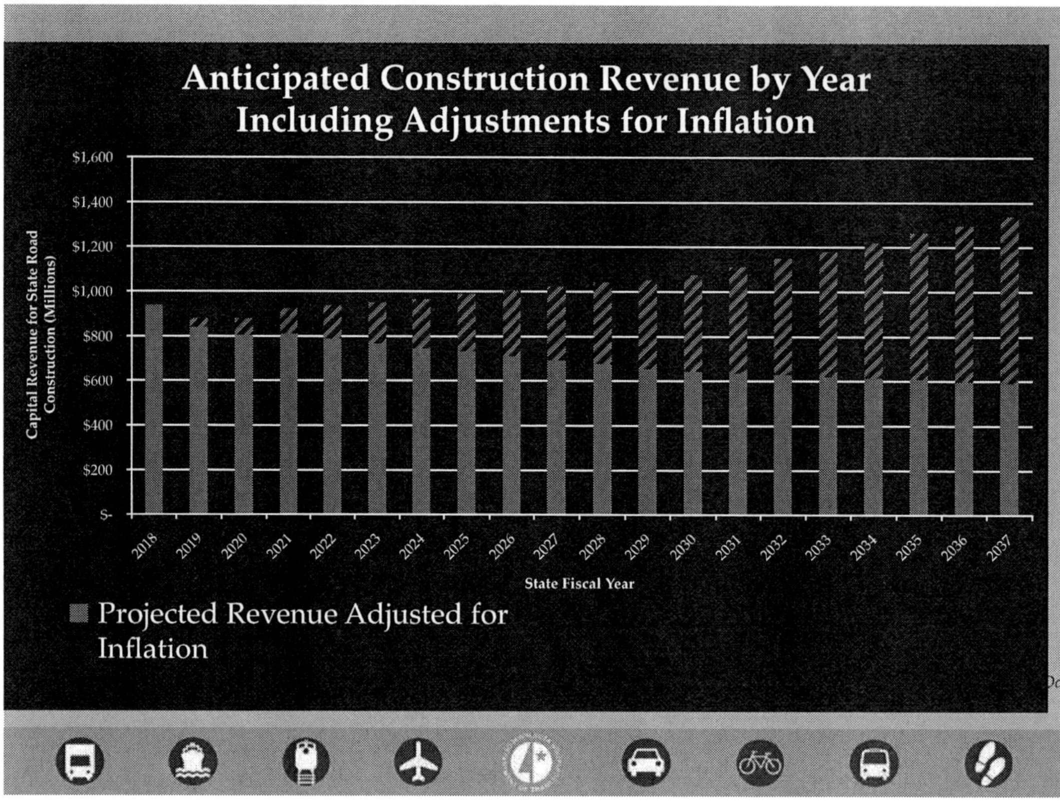


## Construction Projects in Calendar Years 2017-2018 By Program

| Program | Number of Projects | Cost in Millions |
|---|---|---|
| Bridge Replacement | 35 | 143 |
| Major Construction | 7 | 490 |
| Reconstruction | 14 | 119 |
| Recondition | 21 | 123 |
| Resurface | 84 | 381 |
| TOTAL | 161 | 1,257 |

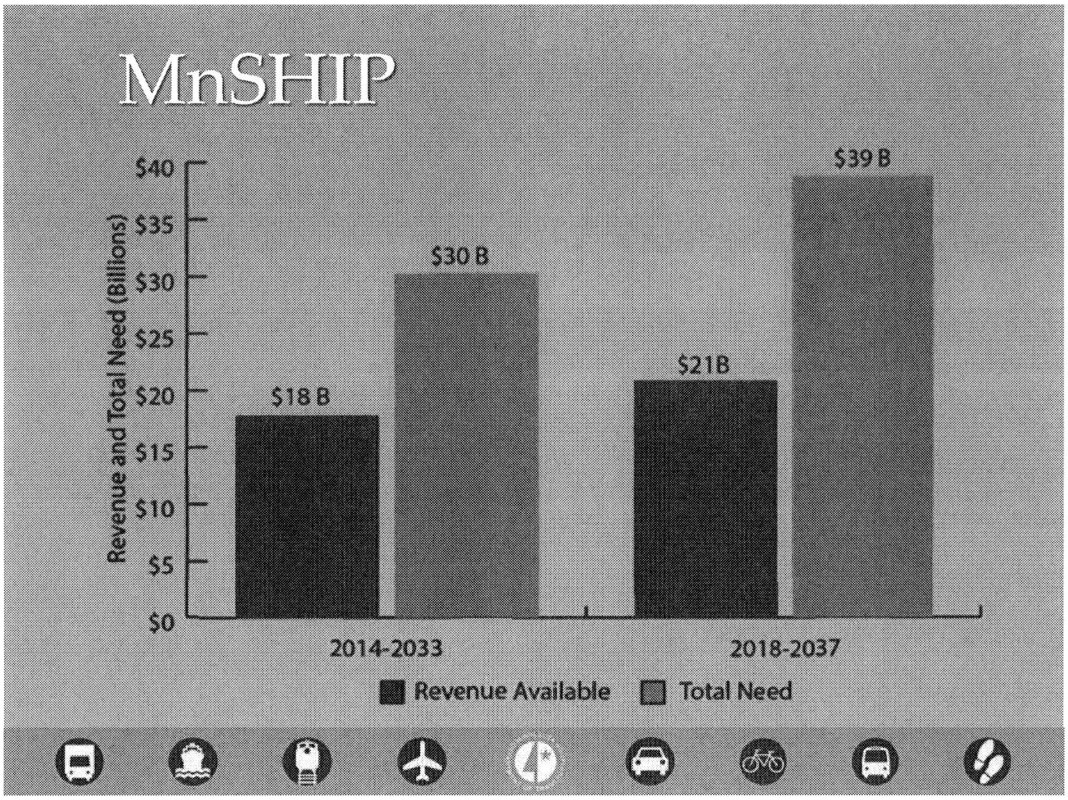

MnSHIP
$40
$35
$30
$25
$20
$15
$10
$5
$0
Revenue and Total Need (Billions)
$18 B
$30 B
$21B
$39 B
2014-2033
2018-2037
Revenue Available
Total Need


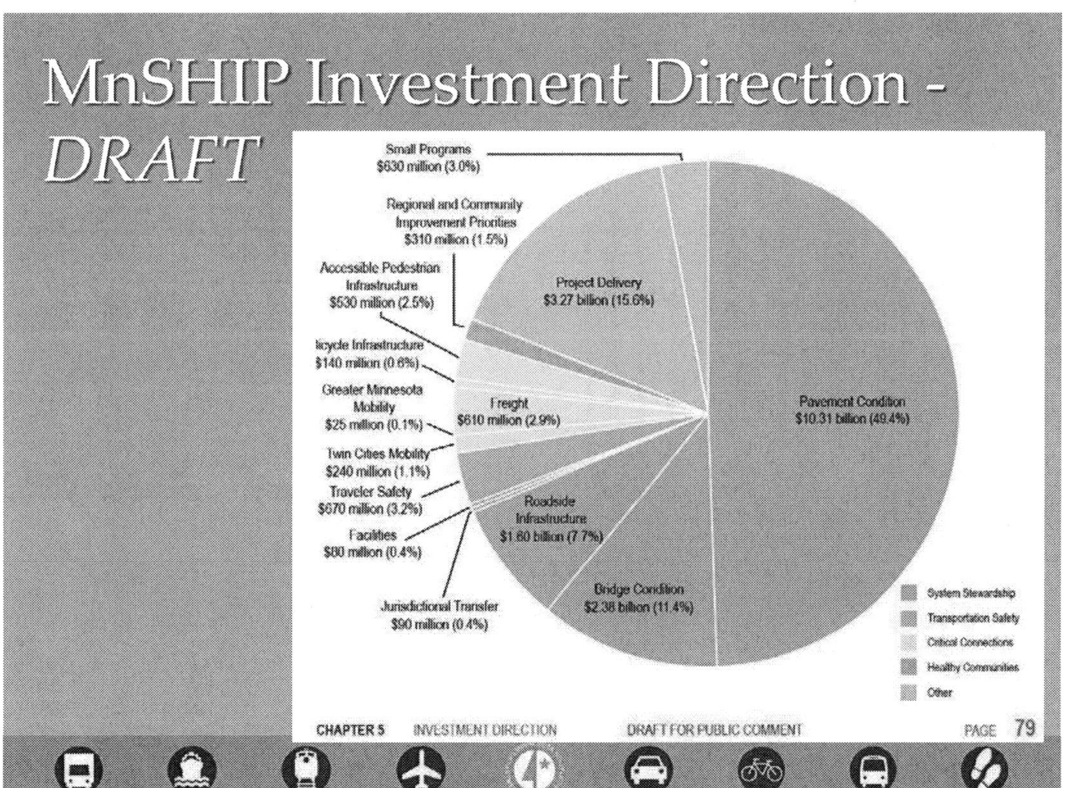

MnSHIP Investment Direction - *DRAFT*
Small Programs $630 million (3.0%)
Regional and Community Improvement Priorities $310 million (1.5%)
Accessible Pedestrian Infrastructure $530 million (2.5%)
licycle Infrastructure $140 million (0.6%)
Greater Minnesota Mobility $25 million (0.1%)
Twin Cities Mobility $240 million (1.1%)
Traveler Safety $670 million (3.2%)
Facilities $80 million (0.4%)
Jurisdictional Transfer $90 million (0.4%)
Project Delivery $3.27 billion (15.6%)
Freight $610 million (2.9%)
Roadside Infrastructure $1.60 billion (7.7%)
Bridge Condition $2.38 billion (11.4%)
Pavement Condition $10.31 billion (49.4%)
System Stewardship
Transportation Safety
Critical Connections
Healthy Communities
Other
CHAPTER 5 INVESTMENT DIRECTION
DRAFT FOR PUBLIC COMMENT
PAGE 79

# Future Corridors of Commerce

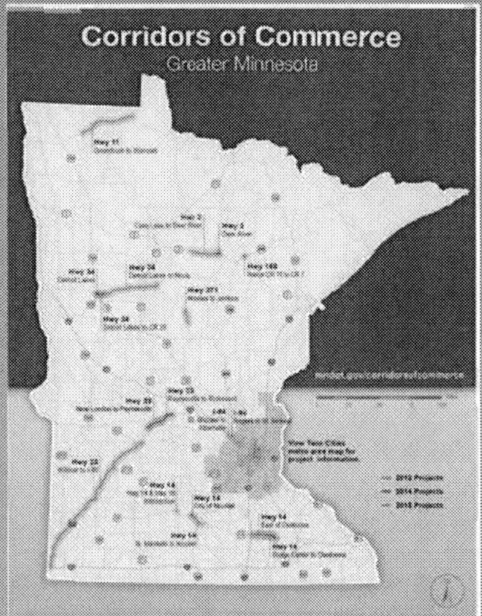


**Projects being readied.**

- Hwy 14 & 15 intersection – New Ulm
- I-94 – St. Michael to Albertville lane addition
- Hwy 11 – Greenbush to Warroad passing lanes
- Hwy 14 – Owatonna to Dodge Center lane addition
- Hwy 23 – New London to Paynesville lane addition
- Hwy 23 - Paynesville to Richmond lane addition
- I-35W – Minnesota River crossing
- I-35W – Northern suburbs MnPASS
- Hwy 65 – Central Avenue bridge deck
- I-94 – Minneapolis and St. Paul

# Transportation Funding –Unfinished Business

- **Transportation was a Top Priority along with Capital Bonding and Tax Bills**
- **All 3 Were Left On the Table**
- **If We Keep Transportation a Top Priority – we have a good shot at getting a bill through the process**

## Funding in Capital Bonding Bill

- Local Bridge Program - $90M
- Local Road Improvement Program - $141M
- Corridors of Commerce -$199.5M
- State and Local Roads Funding - $100.5M
- Trunk Highway Bonds - $62M
- Wetland Mitigation for Local Roads - $5M
- Rail Grade Separation - $26.7M
- Rail Service - $6M
- Port Development Assistance - $5M
- Airports - $14M
- Transportation Economic Development - $7M

## New Legislature in 2017

- GOP Majorities in House and Senate
  - Senate 34-33 (21 new – 15 brand new)
  - House 76-57 (23 New Reps)
- Senate Majority Leader **Paul Gazelka**
- Senate Minority Leader **Tom Bakk**
- House Speaker **Kurt Daudt**
- House Majority Leader **Joyce Peppin**
- House Minority Leader **Melissa Hortman**

## New Legislature in 2017

- Senate Transportation Committee Chair
  - Senator **Scott Newman** (Hutchinson)

  Senate Capital Investment Committee Chair

  **David Senjem** (Rochester)

- House Transportation Committee Chairs
  - Rep. **Paul Torkelson** (Hanska)
  - Rep. **Linda Runbeck** (Circle Pines)

  House Capital Investment Committee Chair

  Rep. **Dean Urdahl** (Grove City)

## Top Issues in 2017

- **Health Care**
- **Tax Relief**
- **Transportation/Infrastructure**

THE TRANSPORTATION ALLIANCE

Highways • Transit • Rail • Waterways • Air

- As a Member of the Transportation Alliance you can join a diverse coalition of organizations all working together to advocate for increased Transportation Funding.

www.transportationalliance.com

# AUTHOR INDEX